Abdelhafid Mimouni

Gout and Antioxidants: The Invisible Antagonists

Abdelhafid Mimouni

Gout and Antioxidants: The Invisible Antagonists

ScienciaScripts

Imprint
Any brand names and product names mentioned in this book are subject to trademark, brand or patent protection and are trademarks or registered trademarks of their respective holders. The use of brand names, product names, common names, trade names, product descriptions etc. even without a particular marking in this work is in no way to be construed to mean that such names may be regarded as unrestricted in respect of trademark and brand protection legislation and could thus be used by anyone.

Cover image: www.ingimage.com

This book is a translation from the original published under ISBN 978-620-6-70596-3.

Publisher:
Sciencia Scripts
is a trademark of
Dodo Books Indian Ocean Ltd. and OmniScriptum S.R.L publishing group

120 High Road, East Finchley, London, N2 9ED, United Kingdom
Str. Armeneasca 28/1, office 1, Chisinau MD-2012, Republic of Moldova, Europe
Printed at: see last page
ISBN: 978-620-7-72382-9

Title: "Gout and Antioxidants: The Invisible Antagonists"

Author: Dr. Abdelhafid Mimouni, PhD in Bioinorganic Systems, Université de Paris XII, Créteil, France (1997). An independent researcher, Dr. Mimouni has extensive expertise in the synthesis, characterization and evaluation of organometallic and macromolecular compounds. His technical skills extend to spectroscopy (IR, UV-Visible, Raman, XAS, NMR) and the use of simulation software such as FEFF and EXAFS for Mac. His thesis (1993-1997) focused on the synthesis and characterization of various organometallic and macromolecular compounds. Since 1997, he has published books focusing on spectral characterization methods and bioinorganic systems. Dr. Mimouni is also the author of three publications in international peer-reviewed journals, dealing in particular with the study of vitamin B12 derivatives by X-ray absorption spectroscopy and the use of the F-test to compare different structural models fitted to EXAFS spectra of cobaloximes and vitamin B12 derivatives.

Preface :

"The Gout and Antioxidants: The Invisible Antagonists" is a comprehensive work that explores oxidative stress and its impact on health in depth. The book's chapters offer a detailed insight into the role of uric acid as an antioxidant, its links to oxidative stress and gout, and strategies for modulating these aspects in the context of gout. In addition, new targeted therapies and clinical implications for gout management are discussed, as is the crucial importance of glutathione in safeguarding vitamin B12 during intestinal transit. This book offers a comprehensive and enlightened view of a subject that is crucial to human health.

Book outline :

Introduction.

In a world where health has become a major issue, the fight against chronic diseases has become a priority. At the heart of these concerns, oxidative stress and its ramifications on human health occupy a prominent place. This book, entitled "The Gout and Antioxidants: The Invisible Antagonists", explores in depth the crucial role of antioxidants in disease prevention and treatment, highlighting their mechanisms of action, complex interrelationships and clinical implications.

Chapter 1: Introduction to Oxidative Stress delves into the fundamentals of this phenomenon, defining oxidative stress and highlighting the sources and effects of free radicals. This chapter lays the foundations for understanding the importance of antioxidants in combating this deleterious process.

Chapter 2: The Role of Uric Acid as an Antioxidant explores an often overlooked aspect of uric acid as an endogenous antioxidant, while examining the pathological consequences of its overproduction.

Chapter 3: Linking Uric Acid, Oxidative Stress and Gout delves into the complex interrelationships between uric acid, oxidative stress and gout, shedding light on the molecular mechanisms underlying this disease.

Chapter 4: Strategies for Modulating Oxidative Stress and Uric Acid in Gout offers practical approaches to balancing oxidative stress and uric acid, highlighting the importance of an antioxidant-rich diet in preventing and treating gout.

Chapter 5: Future Perspectives and Clinical Implications looks to the future of antioxidant research, offering promising avenues for the development of targeted therapies and discussing their potential impact on disease prevention and treatment.

Chapter 6: The Crucial Role of Glutathione in Vitamin B12 Protection broadens the perspective by exploring the link between glutathione and vitamin B12, highlighting the importance of this tripeptide in preserving vitamin B12 during intestinal transport.

This book aims to provide readers with an in-depth understanding of antioxidants and their essential role in preserving health, while exploring the clinical and therapeutic implications of this knowledge. In short, "The Gout and Antioxidants: The Invisible Antagonists" offers a comprehensive and informed view of a subject crucial to human health.

Chapter 1: Introduction to oxidative stress.

The concept of oxidative stress, coined by M. Harman in the 1950s, has since attracted increasing attention in medical research. It refers to an imbalance between the production of free radicals and the body's antioxidant defense mechanisms, leading to cellular and tissue damage.

Sources and effects of free radicals

Free radicals are highly reactive molecules produced during normal oxygen metabolism. Although they are essential to many biological processes, their excessive accumulation can cause oxidative damage to nucleic acids, lipids and proteins, contributing to the development of various chronic diseases.

Antioxidant defense mechanisms :

The body has several antioxidant defense mechanisms to counter the damaging effects of free radicals, such as superoxide (O_2 --), hydrogen peroxide (H O_{22}) and the hydroxyl radical (OH-). Among these mechanisms, glutathione plays a central role in neutralizing these reactive species and protecting cells against oxidative damage.

The role of glutathione in the fight against oxidative stress :

Glutathione exerts its antioxidant effects in several ways. Firstly, it acts as a cofactor for certain antioxidant enzymes, notably glutathione peroxidase, which converts hydrogen peroxide into water, protecting cells against oxidative damage. In addition, glutathione can directly neutralize free radicals by providing them with an electron, transforming them into less reactive, less harmful forms.

Interactions between glutathione and other antioxidants :

Glutathione also interacts with other antioxidants, boosting their effectiveness in the fight against oxidative stress. For example, it regenerates vitamin C and vitamin E, two important antioxidants, by supplying them with electrons, enabling them to continue neutralizing free radicals. In addition, glutathione can protect cells against damage caused by heavy metals by forming stable complexes with them and promoting their excretion by the body.

Role of selenium :

Selenium, an essential trace element, is also crucial to the functioning of glutathione peroxidase. As a cofactor of this enzyme, selenium enables glutathione to neutralize free radicals as efficiently as possible. Selenium thus plays an indispensable role in the body's antioxidant defense against oxidative stress.

Interpretation of the oxidative stress index :

The oxidative stress index is a valuable tool for assessing the level of oxidative stress in the body. It is calculated on the basis of blood levels of various markers, including copper/zinc ratio, glutathione peroxidase, thiols and CRPus. A high index indicates significant oxidative stress, which may be associated with an increased risk of chronic disease.

In conclusion, glutathione, in collaboration with selenium, plays an essential role in the fight against oxidative stress by neutralizing free radicals and protecting cells against oxidative damage. Their interaction with other antioxidants further enhances their effectiveness in preventing chronic diseases associated with oxidative stress. A deeper understanding of these mechanisms may pave the way for new therapeutic strategies to treat and prevent these diseases.

Future prospects :

Research continues to explore new avenues for optimizing the management of oxidative stress and its consequences on health. In-depth studies are needed to fully understand the underlying molecular mechanisms, as well as to identify new therapeutic agents or more targeted prevention approaches. Promising areas of research include techniques to increase glutathione bioavailability, such as the administration of glutathione precursors or the use of compounds that activate glutathione synthesis pathways. In addition, exploring the complex interactions between glutathione, selenium and other antioxidants could reveal new therapeutic synergies.

Clinical implications :

A better understanding of the role of glutathione and selenium in combating oxidative stress may have important clinical implications. Clinicians could consider glutathione or selenium supplementation strategies in patients at increased risk of oxidative stress, such as those suffering from chronic diseases, metabolic disorders or exposed to toxic environments. In addition, monitoring the oxidative stress index could become a useful screening tool to identify individuals at risk and guide preventive interventions.

In short, glutathione and selenium play essential roles in the body's antioxidant defense against oxidative stress. Their complex interaction with other antioxidants and their involvement in a multitude of biological processes underline the importance of maintaining an adequate balance to preserve health and prevent chronic disease.

Natural modulation of glutathione and selenium: Diet, lifestyle and stress management :

Glutathione and selenium modulation can also be addressed through natural approaches such as diet and lifestyle. Indeed, certain foods are rich in glutathione and selenium, and their inclusion in the diet can help increase the body's levels of these antioxidants.

For example, foods such as fresh fruit and vegetables, especially garlic, spinach, avocados and asparagus, are known for their high glutathione content. Brazil nuts, sunflower seeds, tuna, salmon and eggs are also good sources of selenium. By including these foods in the daily diet, it is possible to support the production and action of glutathione and selenium in the body.

In addition to diet, practices to avoid environmental stressors such as pollution, smoking and excessive exposure to UV rays can help prevent the decline in levels of these antioxidants.

Ultimately, a holistic approach to health, including a balanced diet, stress management practices and attention to the environment, can help maintain optimal levels of glutathione and selenium in the body, strengthening the body's ability to fight oxidative stress and prevent associated chronic diseases.

Chapter 2: Role of uric acid as an antioxidant.

Uric acid is often associated with gout, a joint disease characterized by deposits of urate crystals in the joints, but it is also an important endogenous antioxidant in the body. In this chapter, we explore uric acid's role as an antioxidant, its mechanisms of action in neutralizing free radicals, and the pathological consequences of its overproduction.

Introducing uric acid as an endogenous antioxidant :

Uric acid is the end product of purine metabolism in humans. It is synthesized mainly in the liver, where purines from the diet or from cell degradation are converted into uric acid by a series of enzymes. Contrary to popular belief, uric acid is not only a metabolic waste product to be eliminated, but also plays an important role as an antioxidant in the body.

Mechanisms of action to neutralize free radicals :

Uric acid acts as an antioxidant, neutralizing free radicals such as the superoxide radical (O2--) and the hydroxyl radical (OH-). It does this by providing electrons to free radicals, transforming them into less reactive, less harmful species. More specifically, uric acid reacts with the superoxide radical to form water peroxide (H O_{22}), which is then converted to water (H_2 O) by enzymes such as catalase and glutathione peroxidase.

In addition to its ability to neutralize free radicals, uric acid can also protect lipids from oxidation, reduce oxidative stress induced by ischemia and reperfusion, and modulate inflammation. These beneficial effects make it an important component of the body's antioxidant system.

Pathological consequences of uric acid overproduction :

Although uric acid is an important antioxidant, its overproduction can lead to pathological consequences. Due to its inability to regenerate, excessive uric acid accumulation can lead to increased oxidative stress and cell damage. In addition, high levels of uric acid can promote the formation of urate crystals, which can lead to conditions such as gout and kidney stones.

Interestingly, studies have also suggested a link between high uric acid levels and diseases such as cardiovascular disease, hypertension, type 2 diabetes and obesity, although the precise mechanisms of this association are not yet fully understood.

In conclusion, although uric acid is often associated with pathological conditions such as gout, it also plays an important role as an endogenous antioxidant in the body. However, overproduction of uric acid can lead to pathological consequences due to its inability to regenerate and its ability to promote urate crystal formation. A better understanding of these mechanisms could pave the way for new therapeutic strategies to treat diseases associated with uric acid dysfunction.

The importance of uric acid balance in the body:

The balance of uric acid in the body is crucial to maintaining optimal antioxidant function while avoiding the harmful effects of its overproduction. Recent studies have highlighted the importance of this balance in preventing many chronic diseases and promoting optimal health.

As an endogenous antioxidant, uric acid plays a key role in protecting against oxidative stress and inflammation. However, excessively high levels of uric acid can lead to deleterious effects, including cell damage, urate crystal formation and an increased risk of cardiovascular disease, hypertension, type 2 diabetes and obesity.

Thus, maintaining a proper balance of uric acid in the body is essential to prevent these complications. Approaches such as a balanced diet, reducing consumption of purine-rich foods, promoting adequate hydration, and regular physical activity can help maintain appropriate uric acid levels.

In addition, therapeutic strategies aimed at regulating uric acid production and elimination, such as the use of drugs and targeted therapies, could offer new perspectives in the prevention and treatment of diseases associated with uric acid imbalance.

In conclusion, understanding the importance of maintaining an adequate balance of uric acid in the body is essential for preserving health and preventing chronic disease. By exploring the mechanisms by which uric acid is regulated and developing targeted therapeutic approaches, it is possible to optimize its role as an antioxidant while minimizing its harmful effects, paving the way for new health management strategies.

Chapter 3: Link between uric acid, oxidative stress and gout.

Gout, a form of inflammatory arthritis caused by urate crystal deposits in the joints, is closely linked to a high concentration of uric acid in the body. In this chapter, we'll explore in detail how excess uric acid can contribute to oxidative stress and aggravate gout disease, as well as the underlying mechanisms that promote urate crystal formation and trigger an inflammatory response in the joints.

Contribution of uric acid to oxidative stress :

High levels of uric acid in the body can contribute to oxidative stress in several ways. Firstly, uric acid itself can act as an oxidant when present at high levels, causing oxidative damage to tissues and cells. In addition, studies have shown that uric acid can stimulate the production of free radicals, such as the superoxide radical (O2--), by inflammatory cells, thereby increasing the oxidative load in the body. As a result, uric acid-induced oxidative stress can contribute to inflammation and the progression of gout.

Mechanisms of urate crystal formation and joint inflammation :

Excess uric acid promotes the formation of urate crystals, which are solid deposits of sodium urate, in the joints. These crystals usually form when uric acid levels in the blood exceed maximum solubility, causing them to precipitate and deposit in joint tissues. Once formed, urate crystals trigger an acute inflammatory reaction by activating the immune system and recruiting inflammatory cells such as macrophages and neutrophils. These cells then release inflammatory mediators such as interleukin-1β (IL-1β) and tumor necrosis factor alpha (TNF-α), which contribute to inflammation and tissue damage in the joints.

Consequences of joint inflammation in gout :

The joint inflammation associated with gout can cause intense pain, swelling and joint stiffness, significantly affecting patients' quality of life. Moreover, recurrent episodes of untreated inflammation can lead to permanent joint damage, disability and impaired joint function. Consequently, effective management of uric acid and oxidative stress is crucial to preventing and alleviating gout symptoms, as well as reducing the risk of long-term complications.

In conclusion, although uric acid contributes 60% of the total antioxidant capacity of plasma, it is not an antioxidant in the traditional sense, as it cannot regenerate when oxidized. However, its own antioxidant power is undeniable. Paradoxically, high blood levels of uric acid are often associated with pathological conditions such as gout, kidney disease and cardiovascular disease, all of which increase the level of oxidative stress in the body. So, while uric acid may have antioxidant effects, its excess can also exacerbate oxidative stress and the pathological conditions that follow. This nuance underlines the importance of a delicate balance in uric acid regulation to maintain health and prevent associated diseases.

"Uric acid and oxidative stress management in gout: implications for joint health."

This chapter provides a detailed overview of the links between uric acid, oxidative stress and gout, highlighting the mechanisms of urate crystal formation, joint inflammation and the long-term consequences of the disease. To complete this analysis, here's an additional text to integrate:

Effective management of uric acid and oxidative stress is of paramount importance in the prevention and treatment of gout, a form of inflammatory arthritis associated with urate crystal deposits in the joints. Excess uric acid in the body can contribute to oxidative stress, promote urate crystal formation and trigger an acute inflammatory reaction, leading to painful symptoms and long-term joint damage.

Understanding the mechanisms underlying these pathological processes is crucial to developing targeted management strategies. Indeed, uric acid, while playing an important antioxidant role, can become oxidative at high levels, contributing to oxidative stress and inflammation. This duality underlines the need for a delicate balance in uric acid regulation to maintain joint health and prevent gout-related complications.

Uric acid management in gout involves not only reducing excessive uric acid levels, but also addressing the oxidative stress induced by this excess. Therapeutic approaches aimed at reducing uric acid production, enhancing its elimination and mitigating oxidative stress may offer promising prospects for managing gout and preserving joint health.

In conclusion, the delicate balance between the antioxidant and oxidant represented by uric acid underlines the need for a holistic approach to the management of gout. By understanding the mechanisms of urate crystal formation, the impact of oxidative stress and the long-term implications of joint inflammation, it is possible to develop innovative therapeutic strategies to alleviate gout symptoms and prevent permanent joint damage, thereby improving quality of life.

Chapter 4: Strategies for modulating oxidative stress and uric acid in gout.

The management of oxidative stress and uric acid is crucial in the prevention and treatment of gout. Strategies aimed at balancing these two factors may offer promising prospects for alleviating symptoms and reducing the risk of complications associated with this condition.

Modulation of oxidative stress :

To reduce oxidative stress, it's essential to eat a diet rich in antioxidants. Cruciferous vegetables such as broccoli, citrus fruits, berries and colorful fruits and vegetables are all valuable sources of vitamins C and E, selenium and glutathione. These compounds act synergistically to neutralize free radicals and protect cells against oxidative damage.

Glutathione, in particular, deserves special attention. In addition to its ability to neutralize free radicals directly, it also acts as a cofactor for several antioxidant enzymes, strengthening the body's antioxidant defense system. Studies have shown that adequate levels of glutathione can reduce inflammation and improve cardiovascular health, important aspects in the management of gout.

Antioxidant balance :

Maintaining a balance between different antioxidants is crucial to controlling oxidative stress. While glutathione plays a central role, it's important not to overlook the importance of other antioxidants, such as vitamin C, vitamin E and selenium. A balanced diet, including a variety of foods rich in these nutrients, is therefore essential.

However, it's also crucial to note that antioxidant supplementation should not be used excessively, as this could have detrimental effects. A balanced approach, combining a healthy diet with targeted supplementation when necessary, is preferable.

Chapter 4 discusses strategies for modulating oxidative stress and uric acid in gout, highlighting the crucial importance of balancing these two factors to alleviate symptoms and reduce the risk of complications associated with this condition.

The modulation of oxidative stress is presented as a pillar of gout management, emphasizing the adoption of an antioxidant-rich diet. Cruciferous vegetables, citrus fruits, berries and colorful fruits and vegetables are recommended for their content of vitamins C and E, selenium and glutathione, which act synergistically to neutralize free radicals and protect cells against oxidative damage. The essential role of glutathione is emphasized, both for its ability to neutralize free radicals directly and for its action as a cofactor for several antioxidant enzymes, thus strengthening the body's antioxidant defense system.

Antioxidant balance is also highlighted as a key element in modulating oxidative stress, emphasizing the importance of maintaining a balance between different antioxidants such as vitamin C, vitamin E and selenium. The emphasis is on a balanced approach, combining a healthy diet with targeted supplementation where necessary, while cautioning against overuse of antioxidant supplementation.

In sum, this chapter offers practical recommendations for modulating oxidative stress and uric acid in gout, highlighting the importance of a balanced, targeted approach to managing this condition.

Future research :

Future research in this field should focus on the development of targeted therapies aimed at specifically modulating oxidative stress and uric acid in gout. In-depth studies of the mechanisms underlying these pathological processes could pave the way for new, more effective and better-targeted therapeutic approaches. For example, studies on the metabolic pathways involved in uric acid production could lead to the development of drugs that specifically target these processes, thereby reducing uric acid overproduction and alleviating gout symptoms. Clinical implications: The clinical implications of this research are significant, as they could influence future gout prevention and treatment strategies. By better understanding the links between oxidative stress, uric acid and gout, clinicians may be able to propose more personalized and effective interventions for their patients. In addition, early identification of oxidative stress-related risk factors for gout could enable more proactive preventive management, thereby reducing the prevalence and severity of the condition. These advances could also have wider implications for public health, helping to reduce the burden of gout on the healthcare system and improving the quality of life of people with the condition. Finally, a better understanding of the links between oxidative stress, uric acid and gout could pave the way for new strategies to prevent cardiovascular disease, which is often associated with this condition.

Chapter 5: Future prospects and clinical implications.

Future prospects: Future research in gout should focus on the development of targeted therapies aimed at specifically modulating oxidative stress and uric acid. In-depth studies of the mechanisms underlying these pathological processes could pave the way for new, more effective and better-targeted therapeutic approaches. For example, investigations into the metabolic pathways involved in uric acid production could lead to the development of drugs that specifically target these processes, thereby reducing uric acid overproduction and alleviating gout symptoms. Similarly, research into the mechanisms of oxidative stress in gout could lead to the development of therapies aimed at specifically modulating these processes, thus offering new perspectives for the treatment of the disease.

Clinical implications :

The clinical implications of this research are significant, as they could influence gout prevention and treatment strategies in the future. By better understanding the links between oxidative stress, uric acid and gout, clinicians may be able to propose more personalized and effective interventions for their patients. In addition, early identification of oxidative stress-related risk factors for gout could enable more proactive preventive management, thereby reducing the prevalence and severity of the condition. These advances could also have wider implications for public health, helping to reduce the burden of gout on the healthcare system and improving the quality of life of people with the condition. Finally, a better understanding of the links between oxidative stress, uric acid and gout could pave the way for new strategies to prevent cardiovascular disease, which is often associated with this condition. In summary, the future prospects and clinical implications of research into oxidative stress and uric acid in gout offer promising opportunities to improve the management of this disease and reduce its impact on public health.

Chapter 6: The crucial role of glutathione in safeguarding vitamin B12 during intestinal transit.

Introduction: Glutathione (GSH), made up of cysteine, glutamine and glycine, goes beyond its well-known antioxidant role. Its decisive influence in the preservation of vitamin B_{12} on its journey through the intestinal epithelium merits further exploration. This chapter delves into the complex mechanisms of GSH, highlighting the use of EXAFS to examine the molecular links, as well as the specific roles of GSSH and the reduced and oxidized forms of GSH.

GSH in the intestinal epithelium: The intestinal epithelium is a key player in the transport of vitamin B_{12} via transcobalamins I, II and III. This process exposes$_{12}$ vitamin B to oxidative risks.

The saving action of GSH against oxidation:

Abundant in intestinal epithelium, GSH acts as an antioxidant fortress, protecting vitamin B12 from free radicals. Its ability to neutralize oxidants helps maintain vitamin$_{B12}$ in its active state, essential for its biological use.

Interactions between GSH and vitamin B transporters$_{12}$:

Transcobalamins I, II and III, involved in vitamin B12 transport, interact closely with GSH. These interactions regulate vitamin B12 absorption, underlining the crucial influence of GSH.

EXAFS spectroscopy in the analysis of molecular interactions: Studies, including EXAFS analysis, have revealed essential molecular details. This technique revealed that GSH binds to vitamin B12 via a Co-s bond of around 2.21 angstroms, stabilizing the cobalt nucleus against oxidation.

Reduced and oxidized forms of GSH:

GSSH and GSH are the oxidized and reduced forms of glutathione, and influence GSH's ability to protect vitamin B_{12} . The reduced form of GSH is particularly crucial, acting as an electron donor to neutralize free radicals and prevent vitamin B oxidation$_{12}$.

The consequences of GSH deficiency on vitamin B absorption$_{12}$:

Decreased levels of intestinal GSH can compromise the antioxidant defense of vitamin B_{12} , leading to reduced absorption. This disruption can lead to vitamin B deficiency$_{12}$ and metabolic complications.

The influence of nutrition on uric acid production: Appropriate nutrition can modulate uric acid production by controlling purine intake. Foods rich in glutathione, such as broccoli, as well as sources of selenium, vitamin C and vitamin E, can help maintain GSH balance and prevent uric acid overproduction.

In short, by protecting vitamin B_{12} from intestinal oxidation, GSH ensures efficient absorption. Molecular mechanisms, including EXAFS interactions and the influences of reduced and oxidized forms of GSH, open up prospects for the development of innovative therapies against vitamin B12 deficiency. Future research could exploit these interactions to design innovative therapies for metabolic disorders associated with impaired vitamin B12 absorption.

The multifaceted involvement of GSH in vitamin B12 preservation during intestinal transit:

The importance of GSH in vitamin B12 preservation is not limited to its antioxidant role. Indeed, recent studies have also highlighted the link between GSH and the regulation of the expression of genes involved in vitamin B12 metabolism in the intestinal epithelium. This gene regulation maintains a delicate balance between vitamin B12 absorption and metabolism, ensuring an adequate supply of this essential vitamin for many biological functions.

In addition, GSH plays a crucial role in modulating the activity of enzymes involved in vitamin B12 metabolism, notably methylmalonyl-CoA mutase and methionine synthase. These enzymes are essential for the conversion of vitamin B12 into its active forms, and GSH is involved in regulating their activity, thus ensuring proper vitamin B12 metabolism.

Finally, it's important to note that disturbances in GSH metabolism can impact vitamin B12 availability, underscoring the importance of maintaining an optimal GSH balance for efficient vitamin B12 absorption.

In sum, the chapter highlights the multifaceted involvement of GSH in safeguarding vitamin B12 during intestinal transit, going beyond its antioxidant role to include gene regulation, enzyme modulation and metabolic balance. These discoveries open up new prospects for understanding the mechanisms underlying vitamin B12 absorption, and for developing innovative therapeutic strategies aimed at optimizing this absorption.

Conclusion

In conclusion, "Antioxidants: Guardians of Health" offers a deep dive into the fascinating world of antioxidants and their impact on human health. By exploring the scientific basis of oxidative stress, examining the complex relationships between uric acid, oxidative stress and gout, and highlighting the essential role of glutathione in vitamin B protection$_{12}$, this book offers a comprehensive and informed view of the subject.

Throughout these chapters, we have seen that antioxidants are not just molecules that fight free radicals, but play a crucial role in the prevention and treatment of many chronic diseases. By understanding their mechanisms of action and exploring different strategies for modulating oxidative stress, we can look forward to a future in which these natural compounds become powerful tools in the therapeutic arsenal against a wide range of diseases.

However, despite the progress made in this field, many challenges remain. Further research is needed to deepen our understanding of the complex interactions between antioxidants, oxidative stress and human health. In addition, further efforts are needed to translate this knowledge into tangible clinical applications, in order to benefit fully from their therapeutic potential.

Ultimately, "**Gout and Antioxidants: The Invisible Antagonists**" hopes to inspire new research, stimulate thinking and encourage the adoption of lifestyles and therapeutic strategies focused on health promotion and disease prevention. By understanding and valuing the crucial role antioxidants play in our well-being, we can pave the way for healthier, more fulfilling lives for all. In short, this book aims to encourage a holistic approach to health, highlighting the importance of antioxidants in preserving our overall well-being.

Appendix :

Uric acid metabolism :

I) Origin of purine bases Purine bases have three main origins:

- Exogenous: nucleic acids from the diet are degraded into nucleotides by nucleases, then into nucleosides by nucleotidases, and finally into purine bases (adenine or guanine) by nucleosidases.
- Endogenous :
  - Degradation of endogenous nucleic acids during cell renewal or cell lysis.
  - De novo purinosynthesis: takes place mainly in the liver. Ribose-5-phosphate, in the presence of ATP and under the action of phosphoribosyl-pyrophosphate synthetase, yields phosphoribosyl pyrophosphate (PRPP). Transfer of an amine function from glutamine to PRPP by glutamyl phosphoribosyl amino transferase (GPRAT) yields phosphoribosylamine (PRA). A few further steps lead to the formation of inosinic acid, which enables the synthesis of purine bases.

II) Origin of uric acid Uric acid is derived from the degradation of purine bases: adenine and guanine are degraded to inosinic acid, then to hypoxanthine. Through xanthine oxidase, hypoxanthine is converted to xanthine and xanthine to uric acid. Guanine can be directly converted to xanthine via guanase.

III) Elimination of uric acid A) Urinary Uric acid undergoes glomerular filtration, tubular reabsorption in the proximal convoluted tubule and tubular secretion in the distal convoluted tubule.

B) Intestinal Uricolysis of uric acid into allantoin by bacterial flora (presence of uricase).

Lexicon:

1. Nuclease: Enzyme that catalyzes the breakdown of nucleic acids into nucleotides.
2. Nucleotidase: Enzyme that catalyzes the breakdown of nucleotides into nucleosides.
3. Nucleosidase: Enzyme that catalyzes the degradation of nucleosides into purine bases (adenine or guanine).
4. Purinosynthesis de novo: Process of biosynthesis of purine bases from simple precursors such as ribose-5-phosphate.
5. Phosphoribosyl pyrophosphate (PRPP): An important intermediate compound in purine base biosynthesis.
6. Xanthine oxidase: An enzyme involved in the conversion of hypoxanthine to xanthine and xanthine to uric acid.
7. Uricase: A bacterial enzyme that catalyzes the conversion of uric acid into allantoin.
8. Glomerular filtration: Process by which blood is filtered in the glomeruli of the kidneys to form primary urine.
9. Tubular reabsorption: Process by which certain substances are reabsorbed into the bloodstream from the primary urine in the renal tubules.

10. Tubular secretion: Process by which certain substances are excreted in the urine from the blood into the renal tubules.

Printed by Books on Demand GmbH, Norderstedt / Germany